JEAN HEFLIN

PENSTEMONS

The Beautiful Beardtongues of New Mexico

Photographs by Bill Heflin
Drawings by DeWitt Ivey

JACKRABBIT PRESS
Albuquerque, New Mexico
Printed in Korea
Copyright 1997

PARTS OF THE PENSTEMON

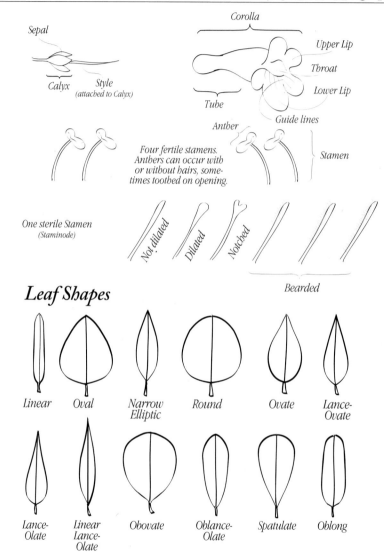

Sepal

Calyx

Style
(attached to Calyx)

Corolla

Upper Lip

Throat

Lower Lip

Tube

Guide lines

Anther

Stamen

Four fertile stamens.
Anthers can occur with
or without hairs, some-
times toothed on opening.

One sterile Stamen
(Staminode)

Not dilated

Dilated

Notched

Bearded

Leaf Shapes

Linear

Oval

Narrow
Elliptic

Round

Ovate

Lance-
Ovate

Lance-
Olate

Linear
Lance-
Olate

Obovate

Oblance-
Olate

Spatulate

Oblong

INTRODUCTION

A number of photographs made when writing with Erma Pilz *The Beautiful Beardtongues of New Mexico; A Field Guide to New Mexico Penstemons* made it seem appropriate to enlarge the collection to include as many photographs of the New Mexico penstemons in the wild as possible. This book is intended to be used by laymen but includes a botanical description (excluding fine details) that hopefully will make it useable for botanists as well.

The almost 300 species of penstemons are members of the Scrophulariaceae (Figwort) family. Their flowers are tube-shaped, ending in five lobes or petals, usually with two at the top and three at the bottom. They have two pairs of fertile stamens, but the enlarged fifth stamen or "staminode" is their most distinguishing mark. The staminode often is bearded, resembling a brush, giving the plant its common name, "Beardtongue". Many penstemons form a basal mat of leaves from which the flower stalks grow and which may disappear at blooming. Others retain this mat. Still others are somewhat bushy or shrubby in habit.

The penstemons appear to be a rapidly evolving genus and are extremely variable, even within one species, sometimes making identification difficult. The height of plants and the number of flowers often will be quite different from a dry year to a wet year. Crossbreeding between species in the wild can occur. I have tried to present the characteristics of the New Mexico species in a way that the layman can make an identification; however, some characteristics, such as anther shape, hairiness or presence of glandular hairs, often may be seen only with a magnifying lens or microscope.

Enjoy the penstemons in the wild; they are one of New Mexico's treasures!

ACKNOWLEDGEMENTS

The author is grateful to the many friends who have given assistance in the preparation of this book, especially to Robert DeWitt Ivey for his drawings and to Bill Heflin who helped me explore the by-ways of New Mexico to find and photograph the penstemons. Carolyn Dodson, David Smoker, Mimi Hubby, Karen Lightfoot, Erma Pilz, Bob Sivinski, Dr. Thomas Todsen, and Ellen Wilde were all especially helpful. Judy Dain helped with the field work and many other friends were supportive and helpful over the years of the project. Thanks as well to Valley and Ed Lowrance and outstanding graphics artist Ronald Jacob for their help in getting the book to press.

Table of Contents

ISBN NO. 0-9659693-0-4

PENSTEMON ALAMOSENSIS

(Alamo penstemon)
Penstemon alamosensis Pennel and Nisbet, Univ. Kans. Sci. Bull. 41:709, 1960

The brilliant scarlet blooms of *Penstemon alamosensis* make it a popular garden plant. Its natural habitat is restricted to the hot, steep limestone western canyons of the Sacramento and the San Andres Mountains. Some years ago it was collected as a landscape plant but its small, scattered populations are fragile and it is now listed as one of the New Mexico Rare and Sensitive Plant Species. Occasionally, it is still found in commercial nurseries though no longer collected in the wild. It blooms from April to June.

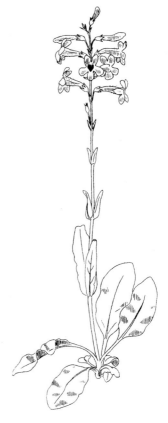

Height and Growth Habit: The height ranges from 12-28" (30-70 cm.) with one or several upright stems arising from a basal rosette. The leaves are grayish or gray-blue in color.

Flowers: The flowers are 3/4-1" (20-25 mm.), bright coral red, with an expanding tube and an almost symmetrical face. The long, narrow, nearly one-sided flowering stem has 1-4 flowers. Side stems, calyces and outside of corolla are sparsely covered with glandular-tipped hairs.

Leaves: Basal leaves are elliptic, obovate or broadly lanceolate, usually stemmed, sometimes blunt tipped, sometimes pointed. Stem leaves are much smaller, in pairs, without stems, pointed, lanceolate or narrowly oblong.

Calyx: The lobes are .12-.2" (3-5 mm.) and are ovate to lanceolate, pointed, with narrow papery edges.

Stamens: The fertile anthers and staminode lie inside the throat; the anthers are flattened; the staminode is hairless.

Penstemon alamosensis
Color Plates Nos. 1 & 2

PENSTEMON ALBIDUS

(Redline penstemon, White flower penstemon)
Penstemon albidus Nutt. Gen. N. Amer. Pl. 2:53 1818

Penstemon albidus is found on the plains and hills of eastern New Mexico from May to July. The pale flowers, white or pale lavender or pink, have black anthers which at first appear like black insects in the flowers. It may be found with dark red guidelines, hence the name Redline Penstemon.

Height and Growth Habit: One or several downy or glandular-hairy stems growing in tufts or clumps reach a height of 8-16" (20-40 cm.).

Flowers: The flowers are .6-.88"(15-22 mm.), pale white, pink or lavender. The inflorescence is narrow, glandular-hairy inside and out, bearing several flowers crowded around the stem.

Leaves: The leaves are dull green, downy or rough, with or without fine teeth at the edges. The basal leaves are stemmed, spatulate or oblong, usually tapered at both ends. The stem leaves are lanceolate.

Calyx: The calyx is .28-.4" (7-10 mm.) long, the lobes lanceolate, sharp-tipped.

Stamens: The narrow staminode has a few hairs at the tip and does not project from the throat. The anthers are black and flattened.

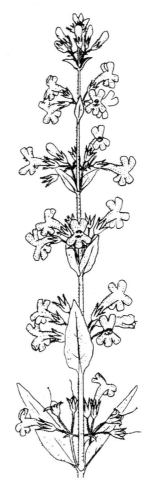

Penstemon albidus
Color Plate No. 3

PENSTEMON AMBIGUUS

(Sand penstemon, Bush penstemon, Moth penstemon, Pink plains penstemon, Phlox penstemon, Gilia penstemon)

Penstemon ambiguus Torr. ssp. *ambiguus*, Ann. Lyc. Nat. His. NY 2:228 1828. ***

Leiostemon ambiguus Greene, Leaflets 1:223 1906

Penstemon ambiguus is one of the most beautiful of New Mexico wildflowers. It is an airy, many branched, shrubby plant growing in sandy, open areas in eastern, central and southern New Mexico. From May to August, depending on the year, it produces a mass of delicate pale to dark pink blooms, sometimes with the flower tubes appearing lavender.

Height and Growth Habit: 8-24" (20-60 cm.), but old plants can reach 2.5-3', many branched. Older plants may have a woody base. Stems and leaves usually are covered with very short stiff hairs.

Flowers: The flowers are .6-.96" (15-24 mm.) with a narrow tube. The flower face is always white, with a suede-like texture but not downy, sometimes appearing pale pink because the rose-pink of the back of the petals shines through. Some flowers may be deeper in color than others. The five nearly equal petal lobes are rounded; the face is flat. The narrow opening of the throat is bearded with short hairs at the base of the lobes, running in two lines down the lower side. Guide lines may be present.

Penstemon ambiguus
Color Plates Nos. 4 & 5

Leaves: The linear leaves are sharp tipped and may be smooth or with short stiff hairs.

Calyx: The calyx is .08-.12" (2-3 mm.) long. The segments are pointed, ovate with ragged papery edges.

Stamens: The staminode is borne inside the throat and is smooth, not dilated.

Note: Two subspecies are listed in New Mexico. Ssp. *ambiguus* occurs in central and eastern New Mexico to eastern Colorado & Utah. Ssp. *laevissimus* differs from the more common *Penstemon ambiguus* Torr. ssp. *ambiguus* in having smooth stems and leaves. Its range extends from Socorro County in New Mexico, south into Mexico, southwestern Texas, west to Arizona & Nevada. *Penstemon ambiguus* may hybridize with *Penstemon thurberi* in some places.

PENSTEMON ANGUSTIFOLIUS

(Broadbeard penstemon, Narrow leaf penstemon, Taperleaf beardtongue)
Penstemon angustifolius Nutt. ssp. *caudatus* (Heller) Keck, Jour. Wash. Acad. Sci. 29:490 1939 *** *P. caudatus* Heller *** *P. angustifolius* var. *caudatus* (Heller) Rydb. *** *P. angustifolius* ssp. *venosus* Keck

Penstemon angustifolius ssp. *caudatus* in New Mexico is a stout, smooth, waxy, gray-green plant that carries its showy blooms in May and June in plains and grasslands of northern New Mexico. The plant is variable and there are several subspecies ranging north to the southern Great Plains, which may intergrade where their ranges overlap.

Height and Growth Habit: The plant grows from 8- 20" (20-50 cm.) with single or several stout, erect or, on some plants, somewhat upward-curving stems .

Flowers: The flowers range from .68-.92" (17-23 mm.) and are pale to royal blue, lavender or pink. The dense flower clusters are tiered and crowded on the stem. The flowers are usually hairless but may occasionally have a few short hairs at the base of the lower lip. Guidelines may be present.

Leaves: The leaves are gray-green and slightly waxy. No basal mat is present at blooming. The basal leaves are lanceolate, oblanceolate, or spatulate with short stems. The middle and upper stem leaves are not stemmed and are broadly lanceolate and sharp tipped.

Calyx: The .16-.28" (4-7 mm.) sepals are very sharply tipped, papery only at the base, lanceolate or lanceovate in shape.

Stamens: The staminode is bearded with yellow hairs, projecting only slightly out of the throat and is dilated at the end.

Penstemon angustifolius
Color Plates Nos. 6 & 7

Note: *Penstemon angustifolius* var. *venosus* grows in San Juan Co. as well as in Arizona and Utah. It is distinguished by its pink-lavender flowers, herbage not darkening on drying and bracts that are strongly veined on both sides. In southern Colorado, *Penstemon angustifolius* ssp. *caudatus*, may intergrade with *ssp. angustifolius*.

PENSTEMON AURIBERBIS

(Golden beard penstemon, Golden beardtongue)
Penstemon auriberbis Pennell, Contr. U.S. Nat. Herb. 20:339, 1920

Penstemon auriberbis is a small plant that is most often found in Colorado but occurs occasionally in sagebrush land and plains of Colfax and Union Counties near the Colorado border. One of its most noticeable characteristics is its golden-bearded staminode which is markedly visible in the mouth of the blooms which occur during May and June. It is listed by the State as one of the New Mexico Rare and Sensitive Plant Species.

Height and Growth Habit: The 4-13.6" (10-35 cm.), hairy plant sometimes occurs with red stems.

Flowers: The blooming spike is narrow, one-sided (or nearly so) with .72-1" (18-25 mm.) lavender or purplish flowers. The flower throat is pale-colored and wide open. The lower lip has a few sparse hairs. The outside of the flower is glandular. The bracts are leaf-like and pointed, the lower bracts being wider than the leaves.

Leaves: The leaves are linear or lanceolate and may be downy or smooth. The basal leaves are narrow, stemmed with smooth or wavy and toothed edges and are present at blooming. The upper stem leaves are usually wider than the basal leaves.

Calyx: The sepals are long .28-.36" (7-9 mm.), lanceolate and pointed.

Stamens: The staminode is bearded to the base with long, bright golden hairs.

Penstemon auriberis
Color Plates Nos. 8 & 9

PENSTEMON BARBATUS

(Scarlet penstemon, Scarlet bugler, Golden beard penstemon, Southwestern penstemon)

Penstemon barbatus is common from early June to frost throughout New Mexico's pinyon-juniper and oak woodlands up to spruce elevations. It is a tall, upright plant with showy, somewhat pendant, bright red tubular flowers, which resemble to a shark's head in profile. The stemmed leaves of the basal mat are present at flowering. Hybrids are known to exist with pink, salmon or pink-lavender flowers, probably crossed with *Penstemon virgatus*. The three subspecies known in New Mexico are all very similar.

❖❖

Penstemon barbatus (Cav.) Roth ssp. *barbatus*, Cat. Bot. 3:49 1806 *** *Chelone barbata* Cav., Icones Pl. 3:22, 242 1794; *** *Penstemon barbatus* var. *puberulous* A. Gray, U.S. and Mex. Bound. Bot. Rpt. 114 1859.

Height and growth habit: The plants occur up to 3' (90 cm.), occasionally taller, with one or several slender, sometimes reddish stems arising from a basal mat. The stems can be smooth or carry short hairs at the base. The plants appear open and airy. The flowering stalks are usually more or less one-sided, although occasionally a more spread-out form is found with flowers occurring on all sides of the main stem.

Flowers: The narrow, smooth, orange-red to scarlet tubes are 1-1.5" (25-38 mm.) long, gradually broadening toward the lips. The upper two lobes are joined and protrude to form a hood, the lower bent back sharply. A few yellow-white hairs may be present at the base of the lower lip.

Leaves: The leaves are narrow, not abundant. The basal leaves are 2-5" (5-13 cm.) in length, lanceolate, spatulate or ovate, may be smooth or hairy, and are present at blooming time. The stem leaves are linear to lanceolate, and without stems.

Penstemon barbatus ssp. *torreyi*
Color Plates Nos. 10 & 11

Calyx: The calyx is .25-.4" (6-10 mm.) long. The pointed lobes are lanceolate, elliptic or ovate with very narrow papery edges and may be slightly downy.

Stamens: The yellow or whitish fertile anthers are slightly toothed and show plainly beyond the tip of the hood. The staminode is bare and smooth, light in color, slightly dilated and does not extend out of the throat.

Note: The range of subspecies *barbatus* includes southwestern New Mexico and extends into Mexico.

Penstemon barbatus ssp. *torreyi* (Benth.) Keck, Jour. Wash. Acad. Sci 29:491. 1939 *** *Penstemon torreyi* Benth, DC Prod 10:324, 1846. *** *Penstemon barbatus* var. *torreyi* A. Gray, Proc. Amer. Acad. Arts and Sci. 6:59, 1862.

Subspecies *torreyi* is more common in central and northern New Mexico mountains. It has more slender stems, less ample foliage with linear stem leaves, and a shorter calyx .12-.2" (3-5 mm.). There is some evidence that this subspecies may hybridize with ssp. *barbatus.*

Penstemon barbatus ssp. *trichander* (A. Gray) Keck, Jour. Wash. Acad. Sci. 29:491. 1939. *** *Penstemon barbatus* var. *trichander* A. Gray, Proc. Amer. Acad. Arts and Sc. 11:94. 1876 *** *Penstemon trichander* Gray Rydb., bull. Torr. Bot. Club 33:151 1906.

Occurring in northwestern New Mexico, subspecies *trichander* is distinguished from the two above by long white hairs on the anthers (sometimes sparse). There is indication that this subspecies has intermixed with *P. strictus* where their ranges overlap.

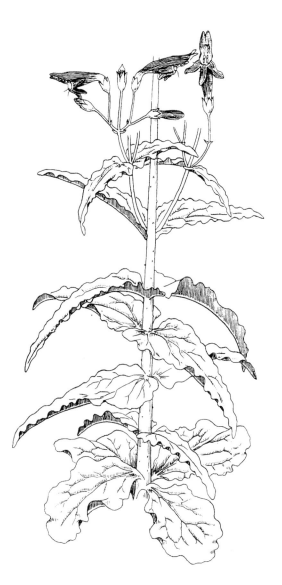

Penstemon barbatus ssp. *barbatus*

PENSTEMON BREVICULUS

(Narrow-mouth penstemon)

Penstemon breviculus (Keck) Nisbet and Jackson, Univ. Kans. Sci. Bull. 41:734. 1960 ***
Penstemon jamesii Benth. ssp. *breviculus* Keck, Bull. Torr Bot. Club 65:241, 1938 (*Baker, Earle & Tracey 410*, dry plains below Mancos, Montezuma Co., Colo., 8 July 1898; holotype at POM! isotype at NY!)

Penstemon breviculus is a very small cousin to *Penstemon jamesii* but is now considered a separate species. It has a bearded but not glandular throat. The staminode lies inside the throat or may protrude slightly and is bearded its full length with the hairs pointing back down the throat. It blooms in May or June in San Juan County. It is listed on the New Mexico List of Rare and Sensitive Plant Species.

Height and Growth Habit: The plant grows only 4" to about 9" (10-25 cm.) with downy, erect or slightly upward curving stems coming from woody rootstocks.

Flowers: The .5-.7" (13-18 mm.) flowers are lavender or dark blue to purple, tubular, funnelform with prominent dark guidelines and are glandular-hairy on the outside. The lower lip is bearded with long yellow-white hairs and is shorter than the upper. The shape of the face is rectangular and narrow.

Leaves: The leaves are downy, becoming smooth with age. The stemmed basal leaves are elliptic, spatulate or lanceolate, usually pointed. The stem leaves are lanceolate, occasionally with a few pointed teeth at the edges, clasping the flower stalk.

Calyx: The calyx is .2-.32" (5-8 mm.) long. The lobes are lanceolate, pointed with narrow papery edges on the lower third and are glandular hairy.

Penstemon breviculus
Color Plates Nos. 12 & 13

Stamens: The staminode is narrow, lying within the throat or barely projecting but clearly visible. It is bearded with yellow almost threadlike hairs for most of its length, those at the tip long, loose and spreading, those behind the tip much shorter, straight and pointing backward.

PENSTEMON BUCKLEYI

(Buckley's beardtongue)

Penstemon buckleyi Pennell, Proc. Acad. Nat. Sci. Phila. 73:486. 1921 *** *Penstemon amplexicaulis* Buckley, Proc. Acad. Nat. Sci. Phila. 13:461. 1862

Penstemon buckleyi is a stout gray-green plant of the dry sandy plains and grasslands of southeastern New Mexico, ranging into Texas. It blooms from April to June, carrying its flowers in whorls around the stem. The bracts are rounded or almost heart shaped.

Height and Growth Habit: The plant has single or several stout, smooth, erect, waxy stems, 12"-16" (30-40 cm.) or sometimes taller.

Flowers: The flowers are carried around the main stem in whorls, each whorl with several lavender, pale blue, pink or white flowers approximately .6-.8" (16-20 mm.) long. The bracts are wide, rounded or almost heart-shaped.

Leaves: The leaves are moderately thick, smooth, waxy, gray-green. The basal leaves are stemmed, mostly oblanceolate or blunt and spatulate but occasionally may be lanceolate and sharp-tipped. The stem leaves are lanceolate or ovate, stemless and sharp tipped.

Calyx: The .16-.2" (4-5 mm.) calyx is smooth. The lobes are rounded at the base, with sharp tips and papery edges.

Stamens: The staminode lies within the throat and is lightly bearded with yellowish hairs.

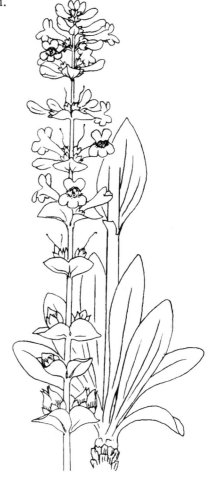

Penstemon buckleyi
Color Plates Nos. 14 & 15

PENSTEMON CARDINALIS

(Cardinal penstemon)

Penstemon cardinalis is noted for its deep red, drooping tubular flowers with a constriction of the tube just behind the flower face. Two subspecies exist and both are on the State of New Mexico Rare and Sensitive Plant Species List.

Penstemon cardinalis Woot. & Standl. ssp. *cardinalis*, Contr. U.S. Nat. Herb. 16:171. 1913 The type was collected by Wooten on White Mountain Peak immediately above the forks of Ruidoso Creek, July 6, 1895. *** *Penstemon crassulus.* Woot. and Standl. Contr. U.S. Nat. Herb. 16:172. 1913

Penstemon cardinalis ssp. *cardinalis (*Scarlet penstemon*)* grows in the Sacramento, Oscuro and Capitan Mountains at pine or spruce-fir elevations. It blooms in May or June.

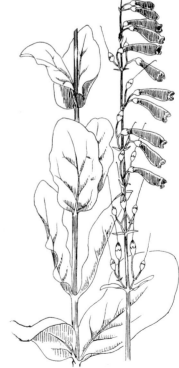

Height and Growth Habit: The plant is often spreading with paired lower leaves which alternate direction 90 degrees as they ascend the stems. There are few to several heavy smooth, green or gray-green stems 15.6-27" (40-70 cm.).

Flowers: The flowers are .88-1.2" (22-30 mm.) long, slightly swollen, dark red or bluish-red tubes, with the same intensity of color throughout. They are carried on a one-sided spike, two blooms to each 1" side stem. The flower face is somewhat compressed. The corolla has small lobes, the upper two erect, the lower 3 spreading and covered with golden hairs which may obscure the guidelines. The constriction around the somewhat closed mouth of the tube makes it appear like a bag with a drawstring partly closed.

Leaves: The leaves are smooth, very gray or bluish-green, waxy and can reach 4.7" long and 2" wide . The basal leaves are short stemmed and blunt at the end, elliptic, spatulate or ovate. The stem leaves are oblong, ovate, lanceolate, pointed at the tip and without stems.

Calyx: The very short calyx .12-.14" (3-3.5 mm.), has ovate lobes which are pointed or blunt, narrowly papery on the edges and which may be red in color.

Penstemon cardinalis
Color Plates Nos. 16, 17 18 & 19

Stamens: All of the stamens lie within the throat. The anthers are toothed on the edge. The staminode is bearded with golden hairs at and near the tip, and is not dilated.

❖❖

Penstemon cardinalis ssp. *regalis* (Guadalupe penstemon, Royal beardtongue) (A. Nels.) Nisbet and Jackson, Univ. Kans. Sci. Bull. 41:707, 1960 *** *Penstemon regalis* A. Nels, Amer. Jour. Bot. 21:578. 1934

P. cardinalis subspecies *regalis* is restricted to rocky ridges of the Guadalupe Mountains in pine, fir or spruce forests. It blooms from May to June. It resembles subspecies *cardinalis* in many respects, especially the flowers. Its leaves are thicker, the lower ones sometimes elliptic. The calyx is longer, .16-.24" (4-6mm.), its lobes being ovate and sharp tipped.

PENSTEMON COBAEA

(Foxglove penstemon, Wild white snapdragon)
Penstemon cobaea Nutt. (P. Hansonii A. Nels) Trans. Amer. Philos. Soc. II 5:182 1837

Penstemon cobaea, common in the Great Plains states and Texas, is not native to New Mexico, but it is spreading rapidly along roadsides and ditches throughout the state. It is a robust penstemon with shiny green leaves, with many large, showy blooms, giving rise to its common names.

Height and Growth Habit: One or several stems may grow to a height of 25" (25-65 cm.), sometimes taller. It has few to many close-growing, heavy, stiffly upright, downy stems. The white hairs of the lower stem point downward. The stems are sometimes colored reddish-purple.

Flowers: The flowers exhibit considerable variations in color and guidelines. In general, the flowers are large 1.5-2.2" (35-55 mm.), broad, puffy and wide-mouthed, with projecting upper lobes and spreading or reflexed lower lobes. The color may be a light purple or rose infused over the outside of the flower, or, occasionally, pure white. The face and interior of the flower are lighter than the exterior. The exterior of the flower may be smooth or hairy. Prominent dark purple guidelines are present at the mouth and extend down the interior of the tube.

Penstemon cobea
Color Plates Nos. 20 & 21

14

Leaves: The leaves are large, thick, opposite, often toothed at the edges, broad, very dark green, hairy on their under surfaces. The young plant has a large rosette of basal leaves, which disappears with age. The basal leaves are oblanceolate or spatulate, smooth, and stemmed. The stem leaves are sessile, usually hairy, lanceolate to lanceovate.

Calyx: The calyx is downy and glandular with lanceolate to lanceovate, sharp-tipped lobes.

Stamens: The staminode protrudes from the throat and is bearded with yellow hairs that point back into the throat.

PENSTEMON COMARRHENUS

(Dusty penstemon)
Penstemon comarrhenus A. Gray, Proc. Amer. Acad. Arts 12:81 1876 (L.F. Ward 462, "slope of the Aquarius Plateau" 8500 ft. Utah 29 July 1875; lectotype by Pennell Contr. U.S. Natl. Herb. 20:356, 1920 at GH!: isolectotype at NY!)

Penstemon comarrhenus is found in gravelly or sandy soil in sagebrush lands associated with pinyon-juniper, oak and Ponderosa pine. It is a delicate plant with an open inflorescence of very pale pink, blue or blue-lavender flowers, often with a pink tube with a blue flower face. It blooms in June or July.

Height and Growth Habit: Up to 2 or 3' (40-90 cm.) stems arise from a basal mat, which may be present when flowering. Two forms exist: one, a narrow, tall spike more or less one-sided; the other is more spreading.

Flowers: The flowers are 1-1.4" (25-35 mm.), pale pink or pale blue or blue-lavender, often with a pink tube and blue face, and carry red guidelines. The tube is puffy on the lower side, the upper lip is erect, the lower lip is spreading. The flower is smooth inside and out. The flowering stalk is often somewhat one-sided with several rather long side branches. There are several flowers to each cluster, usually at least two in bloom at the same time.

Penstemon comarrhenus
Color Plates Nos. 22 & 23

Leaves: The leaves are dark green. The basal leaves are stemmed, oblong to lanceolate. The stem leaves clasp the stem and are linear to lanceolate.

Calyx: The .14-.24" (3.5-6 mm.) lobes of the calyx are smooth to somewhat hairy, ovate, blunt to sharp tipped, with ragged papery edges.

Stamens: The staminode, which lies within the throat, is smooth, white, dilated and notched at the tip, occasionally with a few hairs at the tip. The fertile anthers protrude and are covered with long, soft, white hairs.

PENSTEMON CRANDALLII SSP. GLABRESCENS

(Crandall's penstemon)
Penstemon crandallii A. Nels ssp. *glabrescens* (Pennell) Keck, Bull. Torrey Bot. Club 64:369. 1937 *** *Penstemon glabrescens* Pennell, Contr. U.S. Nat. Herb. 20:375 1920

Penstemon crandallii ssp. *glabrescens* grows in open areas of northcentral New Mexico's pinyon-juniper and pine woodlands, sometimes in spruce and fir woods. In June, July or August, this ground-hugging, tufted plant is covered with clear blue or lavender flowers on the top half of its stems. It has small, narrow, pointed leaves.

Height and Growth Habit: The plants have many 3.5-9.75" (9-25 cm.) stems that as they grow older and woody may root where they touch the ground. The stems are covered with hairs that may be erect or point backwards.

Flowers: The plant has many one-sided flowering stems that carry .68-.92" (17-23 mm.) wide open flowers on each side stem. The lobes are clear, often sky, blue. The inside of the tube is pale, appearing almost white. The lower lip of the flower has a few scattered hairs, and the outer part of the flower is hairy. Guidelines may be present.

Leaves: The .4-1.4" (10-35 mm.) leaves are dark green, linear, needle-like, sharp tipped, downy at the base and along a middle line of the underside or smooth.

Calyx: The lobes are long and pointed or with a long thin tip, ovate, papery and ragged, .2-.32" (5-8 mm.) long.

Penstemon crandallii
Color Plates Nos. 24 & 25

16

Note: *Penstemon crandallii* ssp. *glabrescens* var. *taosensis* is identical to subspecies *glabrescens* except the leaves are hairy on both sides with fine erect or backward pointing hairs. It is apparently confined to Taos County and nearby Rio Arriba County. It is often intermixed with the *P. crandallii* ssp. *glabrescens*. Several other subspecies grow in nearby Colorado.

PENSTEMON DASYPHYLLUS

(Grama grass penstemon, Gila beardtongue)
Penstemon dasyphyllus A. Gray, U.S. and Mex. Bound. Bot. Rpt. 112. 1859 The type was collected at Cook's Spring by Wright in 1859.

When not in bloom, *Penstemon dasyphyllus* blends in so closely with the grama grass and other grey desert plants that it is almost invisible. More often found in Arizona, Texas and south to Mexico, it is at the edge of its range in Hidalgo and Luna counties, where occasionally it may be located.

Height and Growth Habit: The plant grows from 8-16" (20-40 cm.), sending up few to several stems. The stems are covered with gray hairs, or becoming smooth with age.

Flowers: The flowers are 1-1.4" (25-35 mm.), blue or purplish with the lower lobes longer than the upper, carried on an erect, rather narrow, usually one-sided flower stalk covered with glandular hairs. The glandular-hairy flowers are carried on one-flowered side shoots.

Leaves: The leaves are long 1.5-4.7" (4-12 cm.), narrow, linear, tapering to a sharp point and may be smooth or hairy, occasionally slightly toothed.

Calyx: The calyx measures .16-.28" (4-7 mm.), the lobes being oblong to oblong-lanceolate, tapering to a point, sometimes with narrow papery edges at the base.

Stamens: The staminode is smooth, not dilated and lies within the throat. The anthers are slightly toothed on the edges.

Penstemon dasyphyllus
Color Plates Nos. 26 & 27

PENSTEMON EATONII

(Scarlet bugler, Firecracker penstemon, Eaton's penstemon)
Penstemon eatonii A. Gray, Proc. Amer. Acad. Arts and Sci. 8:395 1872

Penstemon eatonii has exceptionally deep green, glossy leaves that contrast vividly with its showy scarlet tube flowers. Blooming in May or June, it occurs occasionally in northwestern New Mexico.

Height and Growth Habit: *Penstemon eatonii* can grow from about 12-24" (30-60 cm.) or taller. One to several smooth or hairy stems arise from a leafy crown.

Flowers: The flowers are deep scarlet red, nearly straight tubes, .8-1.2" (20-30 mm.), carried on a narrow, many-flowered, usually one-sided inflorescence. The tube is hairless inside and out. The flower face is rounded. The almost equal lobes not much larger than the diameter of the tube. The side flowering stems and the upper part of the principal flower stem may be dark red.

Leaves: The basal and lower leaves are stemmed, elliptic to broadly obovate. The upper leaves are broadly lanceolate to ovate, pointed without stems.

Calyx: The calyx is .16-.32" (4-8 mm.) in length, with red-green ovate lobes which taper to sharp tips, with papery, ragged edges.

Stamens: The staminode lies within the tube and may be reddish in color. It is smooth or slightly bearded at the tip. The fertile anthers may be either inside the throat or project slightly out of the throat and are toothed.

Penstemon eatonii
Color Plates Nos. 28 & 29

Note: It is probable that it is *P. eatonii* ssp. *undosus* which has hairy foliage, that is present in New Mexico, according to Gladys Nisbet.

PENSTEMON FENDLERI

(Fendler penstemon, Plains penstemon)
Penstemon fendleri Torr. and Gray, U.S. Rpt. Expl. Miss. Pacif. 2:168 1855. The type was collected on the Pecos and Llano Estacado in 1854.

Penstemon fendleri grows in sandy or gravelly open land throughout central, eastern and southern New Mexico, extending into Oklahoma and Texas to southeastern Arizona and northern Chihuahua. Blooming from April in the south to August in the north of its range, it is a handsome plant with showy blue or violet flowers carried in distinct whorls. A distinctive characteristic is its "sway-backed" flower buds.

Height and Growth Habit: The single or several stout stems of this plant grow to 7.8-20" (20-50 cm.).

Flowers: The flowers are .68-1" (17-25 mm.) in length, violet or blue with dark violet guidelines. The flower tubes are narrow, slightly curved, with wide open throats and smooth or with a few scattered white hairs on the lower lip. The flowers occur in distinct whorls around the stalk.

Leaves: The leaves are thick, gray-green, smooth or waxy, with pointed tips. The basal leaves are stemmed, wide at the base, lanceolate, elliptic, narrowly ovate. The stem leaves are lanceolate to ovate.

Calyx: The calyx is .16-.28" (4-7 mm.) long with ovate lobes, with pointed tips and with wide papery margins.

Stamens: The staminode lies inside the throat and is bearded with brownish gold hairs at or near the dilated tip.

Penstemon fendleri
Color Plates Nos. 30 & 31

PENSTEMON GLABER VAR. BRANDEGEI

(Blue penstemon)

Penstemon glaber var. *brandegei* (Porter) Freeman *** *Penstemon alpinus Torr.* ssp. *brandegei* (Porter) Harrington. *** *Penstemon brandegei* Penland, Man. Pl. Colo. 496 1954 *** *Penstemon brandegei* Porter ex Rybd. Mem. N.Y. Bot. Gard 1:343, 1900 *** *Penstemon cyananthus* Hook. var. *brandegei* Porter and Coult., Syn. Fl. Colo. 91 1874

This penstemon until recently carried the name *Penstemon alpinus* ssp. *brandegei*. It is not an alpine plant but grows in gravelly or sandy slopes and roadsides at lower elevations in the northeastern New Mexico mountains and plains. It is a robust plant with many large rich blue flowers crowded on a one-sided flower stalk. It blooms in June or July.

Penstemon glaber
Color Plates Nos. 33 & 34

Height and Growth Habit: This penstemon grows from 12- 24" (30-60 cm.) with several stout stems that are finely hairy below and in the inflorescence. No basal mat of leaves is present at blooming time.

Flowers: Many 1.2-1.6" (30-40 mm.) flowers crowd onto the upper third or half of the one-sided flower stalk. The large bracts continue up into the flowering section of the stem. The flower color is strong and varies from purple to blue-purple to pure turquoise-blue, but the flower buds may be flushed with pink. The lower lip of the flower may be slightly hairy.

Leaves: The leaves are smooth or sometimes slightly rough at the edges, appearing crowded on the stem. The upper stem leaves are elliptic or lanceolate or ovate with pointed tips, stemless or even clasping the stem. The lower leaves are smooth, lanceolate or oblanceolate with short stems.

Calyx: The calyx is .24-.32" (6-8mm.) long. The lobes are ovate or rounded, papery and ragged at the edges and sharp tipped.

Stamens: The staminode is smooth or occasionally has a few hairs on its notched tip, and projects out of the throat. The anthers have a few stiff hairs and are toothed on the edges.

PENSTEMON GRACILIS

(Slender beardtongue)

Penstemon gracilis Nutt., gen. N. Amer. Pl. 2:52. 1818 *** *Penstemon pubescens* var. *gracilis*
A. Gray, Proc. Amer. Acad. Arts and Sci. 6:69. 1862-63

Penstemon gracilis is not one of the showy penstemons. It is a plant of the plains and prairies and is at the southern limit of its range in New Mexico where it is occasionally found in open areas in the lower mountains of northeastern New Mexico. It is a branching plant with toothed, dark green leaves and pale, rather flattened flowers in June or July.

Height and Growth Habit: The plant has single or few hairy stems that rise to 8-20" (20-50 cm).

Flowers: The small, .6-.8" (15-20 mm.) flowers are pale blue-violet or whitish lavender and appear flattened but the flower opens widely. The outside of the flower is glandular-hairy and is suffused with pale purple which deepens to almost wine at the juncture with the calyx. The inside of the flower is lighter in color, sometimes almost white. The narrow throat has two ridges on the lower side. The lower lip projects outward and seems larger than the upper which may curl back. White hairs appear at the base of the lower lip. Guidelines may be present. There are usually several blooms on each side stem.

Leaves: The leaves are dull, dark green, smooth or downy. The basal and lower stem leaves have short stems and are oblanceolate, blunt, with or without fine teeth at the edges. The upper stem leaves may be unevenly toothed and without stems, linear or lanceolate and come to a sharp point.

Calyx: The calyx is .16-.2" (4-5 mm.) long, with glandular-pubescent ovate and tapered lobes.

Stamens: The staminode is somewhat dilated, covered with yellow hairs for half to most of its length, visible but not projecting from the throat. The anthers are minutely toothed on the edges, and are almost black-purple.

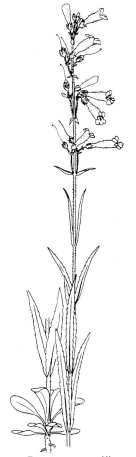

Penstemon gracilis
Color Plate No. 32

PENSTEMON JAMESII

(James' penstemon)

Penstemon jamesii Benth, ssp. *jamesii*, DC Prod. 10:325. 1846. *** *Penstemon similis* A. Nels, bull. Torr. Bot. Club 25:548. 1898

In June and July, *Penstemon jamesii* is one of the joys of our plains and road-sides, often growing in large colonies near the forest edges. This rather short, erect plant is crowded with lavender flowers with clearly visible yellow-bearded staminodes. Note: *Penstemon breviculus* and *Penstemon ophianthus* have at times been considered subspecies of *P. jamesii* but are now classed as species in themselves. See descriptions in this book.

Height and Growth Habit: The plant grows from 4-19" (10-50 cm), depending on moisture available, with one to several, erect, smooth or downy stems.

Flowers: The 1-1.4" (25-35mm) flowers are broad, puffed out above and below and large for the size of the plant. They are crowded onto the one-sided stem. They are lavender or lavender-blue in color with noticeable purple guidelines inside the throat. The large lower lip has numerous long white hairs. The upper lip projects out. The outsides of the flower tubes have glandular hairs.

Leaves: The leaves may be shiny, dark green to almost grey-green, smooth or hairy. Different colonies may have toothed or smooth-edged leaves. The lower leaves are stemmed, linear and tapered at the end, or spatulate and blunt, or lanceolate and tapered at both ends. The stem leaves are linear or lanceolate and taper to a point. The bracts are conspicuous among the flowers and often are longer than the flowers.

Calyx: The calyx is .32-.48" (8-12 mm.) with lanceolate, to ovate lobes, tapered or sharp-tipped, glandular-pubescent, sometimes with papery edges.

Penstemon jamesii
Color Plates Nos. 35 & 36

Stamens: The staminode is narrow with long golden hairs at the tip and shorter hairs behind which point back into the throat. It is is easily seen in the open throat.

PENSTEMON LENTUS

Penstemon lentus Pennell. Contr. U.S. Natl. Herb. 20:359. 1920 (*F. Baker 596*, Arboles, Archuleta Co., Colo. 3 June 1899; holotype at NY!) *** *P. lentus* subsp. *albiflorus* Keck. Amer. Midl. Naturalist 23: 616. 1940. *** *P. lentus* var. *albiflorus* Reveal. Great Basin Naturalist 35:370.1975 (1976). (*C. L. Porter 1801*, Abajo Mts., about 8 mi. w. of Blanding, near the Bear's Ears, 8000 ft. San Juan Co., Utah, 9 June 1939, holotype at RM!, isotype at NY!)

Penstemon lentus has occasionally been reported in the Four Corners area; it is on the New Mexico Rare Plant Review List. It blooms from May to June in dry, sandy or gravelly pinyon-juniper to ponderosa pine elevations. It is a low-growing herbaceous penstemon with blue flowers.

Height and Growth Habit: The plant is usually 8", but sometimes up to 16" (20-40 cm.), in height. One to several erect, smooth and waxy stems arise from a conspicuous rosette of small basal leaves.

Flowers: The flowers are .7-.8" (17-20 mm.), blue to purple, distinctly two-lipped and funnel-form. (Subspecies *albiflorus*, from the Abajo Mountains of southern Utah, is white.) Usually smooth on the outside, they are sometimes bearded with white hairs inside the throat. They are carried on side stems on a somewhat one-sided inflorescence.

Leaves: The leaves are fleshy, blue-green, smooth and waxy. The basal leaves are stemmed, wide and obovate. The stem leaves are lanceolate to ovate with sharp points and without stems.

Calyx: The calyx lobes are tiny .2-.27" (5-6.7 mm), smooth, lanceolate to ovate, blunt or pointed with smooth or ragged, narrow papery edges.

Stamens: The staminode is dilated, reaching the flower opening, and is densely bearded with short, deep yellow hairs near the tip, and more sparsely bearded toward the middle.

PENSTEMON LINARIOIDES

(Linarialeaf penstemon, Toadflax penstemon, Narrowleaf penstemon)

Several subspecies of *Penstemon linarioides* are listed for New Mexico. All are low-growing, spreading plants with very narrow leaves and few to several stems arising from a woody base. They are extremely variable but are excellent ground cover plants, with flowers in the blue to lavender range.

Penstemon linarioides A. Gray ssp. *linarioides*, U.S. and Mex. Bound. Bot Rpt. 112, 1859. The type was collected near the copper mines at Santa Rita, Grant County by Wright.

A low, woody, spreading plant with gray-green pine-like leaves that is found in Catron, Dona Ana, Grant, Hidalgo, Luna, Sierra, and Socorro counties in southwestern New Mexico from late May to August in open grassy areas at pinyon-juniper to pine elevations. The plant is variable; some forms are erect and tall, others are short and compact.

Height and Growth habit: The 8 to 20" (20-50 cm.) stems have fine, appressed or scalelike hairs, which may disappear with age.

Flowers: The flowers are .6-.8" (16-20 mm.), pale to medium blue or blue-purple with a slender purplish-violet tube, with an abruptly expanding throat which may appear pale on the inside. The flowers are carried in a narrow flower cluster turned to one side. They may have deep purple guidelines and are heavily bearded with yellowish hairs at the base of the lower lobes.

Leaves: The leaves are linear .4-1" (10-25 mm.) long, .06-.08" (1.5-2 mm.) wide, with a sharp tooth-like tip, gray-green, usually downy to hairy, crowded near the base of the stems.

Calyx: The .16-.28" (4-7 mm.) lobes are ovate to lanceolate with papery edges , tapering to a point or sharp-tipped.

Stamens: The staminode is bearded for almost all its length with bright yellow hairs, tufted at the tip and visible in the throat.

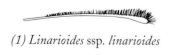

(1) Linarioides ssp. *linarioides*

(2) Linarioides ssp. *coloradoensis*

Penstemon linarioides
Color Plates Nos. 37 & 38

Plate 1
Penstemon alamosensis

Plate 2
Penstemon alamosensis

Plate 4
Penstemon ambiguus

Plate 3
Penstemon albidus (J. Dain photo)

Plate 5
Penstemon ambiguus

Plate 6
Penstemon angustifolius
McKinley County

Plate 7
Penstemon angustifolius

Plate 8
Penstemon auriberis near the Colorado border

Plate 9
Penstemon auriberis

Plate 10
Penstemon barbatus
Jemez Mountains

Plate 11
Penstemon barbatus

Plate 12 Penstemon breviculus
San Juan County

Plate 13 Penstemon breviculus

Plate 14
Penstemon buckleyi
Eddy County

Plate 15
Penstemon buckleyi

Plate 18 Penstemon cardinalis ssp. *regalis*
Guadalupe Mountains

Plate 16 Penstemon cardinalis ssp. *cardinalis*
Capitan Mountain

Plate 17 Penstemon cardinalis ssp. *cardinalis*

Plate 19 Penstemon cardinalis ssp. *regalis*

Plate 20 Penstemon cobea

Plate 21 Penstemon cobea

Plate 22 Penstemon comarrhenus
Near Lindrith, Rio Arriba County

Plate 23 Penstemon comarrhenus

Plate 24 Penstemon crandalli

Plate 25 Penstemon crandalli

Plate 26 Penstemon dasyphyllus
Photographed near Dragoon, Arizona

Plate 28 Penstemon eatonii
Jemez Mountains

Plate 27 Penstemon dasyphyllus

Plate 29 Penstemon eatonii

Plate 30 Penstemon fendlerli

Plate 32 Penstemon gracilis
Garden photo

Plate 31 Penstemon fendlerli

Plate 33 Penstemon glaber var. *brandegii*
Union County

Plate 34 Penstemon glaber var. *brandegii*

Plate 35 Penstemon jamesii
Colfax County

Plate 36 Penstemon jamesii

Plate 37 Penstemon linarioides
Near Bluewater Lake State Park

Plate 38 Penstemon linarioides
(J. Dain photo)

Plate 39
Penstemon neomexicanus
Lincoln National Forest

Plate 40
Penstemon neomexicanus

Plate 41
Penstemon oliganthus
Mount Taylor

Plate 42
Penstemon oliganthus

Plate 43
Penstemon griffini

Plate 44 Penstemon metcalfei
Black Range

Plate 46 Penstemon pseudoparvus
Magdalena Mountains

Plate 45 Penstemon metcalfei

Plate 47 Penstemon pseudoparvus

Plate 48 Penstemon ophianthus
Tijeras Canyon, Sandia Mountains

Plate 49 Penstemon ophianthus

Plate 50 Penstemon palmeri
Garden photo

Plate 51 Penstemon palmeri

Plate 52 Penstemon pinifolius
Near Emory Pass, Black Range

Plate 53 Penstemon pinifolius

Plate 54 Penstemon pseudospectabilis
Catron County

Plate 55 Penstemon pseudospectabilis

Plate 57 Penstemon ramosus

Plate 56 Penstemon ramosus
Photographed near Portal, Arizona

Plate 58 Penstemon rostriflorus
Catron County

Plate 59 Penstemon rostriflorus

Plate 60 Penstemon rydbergi
Near Hopewell Lake, Carson National Forest

Plate 61 Penstemon rydbergi
(D. Ivey photo)

Plate 62 Penstemon secundiflorus
Black Mesa, Rio Arriba County

Plate 63 Penstemon secundiflorus

Plate 64 *Penstemon strictiformus*
Navajo Reservoir, San Juan County

Plate 67 *Penstemon strictus*

Plate 66 *Penstemon strictus*
(E. Pilz photo)

Plate 65 *Penstemon strictiformus*

Plate 68 Penstemon superbus

Plate 69 Penstemon superbus

Plate 70 Penstemon thurberi

Plate 71 Penstemon thurberi

Plate 73 Penstemon virgatus

Plate 72 Penstemon virgatus

Plate 74 Penstemon whippleanus

Penstemon linarioides ssp. *coloradoensis* (A. Nels.) Keck, Bull. Torr. Bot. Club 64:375. 1937. *** *Penstemon coloradoensis* A. Nels., Bull. Torr. Bot. Club 26:355. 1899.

Penstemon linarioides ssp. *coloradoensis* resembles ssp. *linarioides* in many respects but varies in the bearding of the staminode. (See drawing)

Height and Growth Habit: The numerous flowering stems are 4-14" (10-35 cm. tall) and rise from branching rootstocks or older stems to form mats.

Flowers: The corollas are .6-.8" (15-20 mm.) long, lightly bearded at the base of the lower lobes.

Leaves: The leaves are scalelike, downy, sometimes becoming smooth with age.

Calyx: The lobes are ovate to elliptical, tapering or with a point.

Stamens: The staminode is bearded at the tip with a tuft of bright yellow hairs and behind the tuft with sparse white or yellowish hairs.

Penstemon linarioides ssp. *maguirei* Keck, Bull. Torr. Bot. Club 64:378. 1937

First observed in Arizona at a site destroyed by mining, subspecies *maguirei* is known in New Mexico only from an 1880 collection in the Gila River Valley and in the San Francisco River Valley just east of the New Mexico border. It is similar to ssp. *linarioides* except that the leaves are oblanceolate, the lower ones .1-.2" (2.5-5 mm.) wide and blunt at the tip, tapering to the base. It is included on the New Mexico Rare Plant Review List.

Penstemon linarioides ssp. *compactifolius* Keck, Bull. Torr. Bot. Club 64:376. 1937

Recently discovered in the Black Range and Alamo Hueco Mountains above 6000' elevation, *Penstemon linariodes* subspecies *compactifolius* is common in the Flagstaff, Arizona, area. As its name implies, it is a very short plant with tiny, mostly .4" (1 cm.) closely-overlapping leaves. The stems ascend from a ground-hugging rootstock.

PENSTEMON NEOMEXICANUS

(New Mexican penstemon)

Penstemon neomexicanus Woot. and Standl., Contr. U.S. Nat. Herb. 16:172. 1913. The type was collected in pine woods on Eagle Creek in the White Mountains, August 15, 1907, by Wooton and Standley.

There appear to be two population centers for this plant. One is in southern New Mexico in Lincoln and Otero Counties at Sierra Blanca, in the Ruidoso area and the Lincoln Forest at 6000-9000' elevation. The second is in the far north-central section of New Mexico. It is closely related to *Penstemon virgatus* and is similar in appearance but carries its lilac, blue or purple flowers more loosely. Its flowers are larger and it has a bearded lower lip. Its blooming time is from July to August.

Height and Growth Habit: The plant grows from 15-28" (40-70 cm.). Its single or several stems and leaves may be smooth or waxy.

Flowers: The one-sided flower spike carries many 1-1.4" (26-35 mm.) blooms but does not appear crowded. The flower color may be lilac, blue or purple. The throat is broadly expanded and the lower lip is bearded at the base and slightly folded to the back.

Leaves: The leaves are smooth and dark green. The basal leaves are lanceolate or oblanceolate with short stems which may be absent on flowering plants. The stem leaves are lanceolate or linear.

Calyx: The calyx is .16-.28" (4-7 mm.) in length, having wide blunt sepals with ragged papery margins.

Stamens: The staminode is smooth, dilated, often notched at the tip.

Penstemon neomexicanus
Color Plates Nos. 39 & 40

PENSTEMON OLIGANTHUS

Some plants in New Mexico considered at one time to be *Penstemon oliganthus* now have been classified as separate species: *Penstemon griffinii, Penstemon pseudoparvus,* and *Penstemon inflatus. Penstemon metcalfei* is also included in the Oliganthus group. Their descriptions are included below. Studies are underway to clarify the taxonomy of these plants.

(Tubtop penstemon)
Penstemon oliganthus Woot. and Standl., Contr. U.S. Nat. Herb. 16:172. 1913 The type was collected in the mountains west of Grants, Valencia County, August 1, 1892.

Penstemon oliganthus is a delicate plant with small, scattered flowers with a noticeable white throat and white on the lower part of the tube that contrasts with the darker blue or purple of the upper tube and flower face. It grows in forest meadows above 6000' throughout northern New Mexico mountains; for instance, on Mount Taylor, at the base of Antonito Peak or at the top of the Sandia and Manzano Mountains. Occasionally, it can be found in deep shade.

Height and Growth Habit: The plant grows from 6-24" (15-60 cm.), sometimes taller. The single or several stems arise from a basal rosette and are often downy.

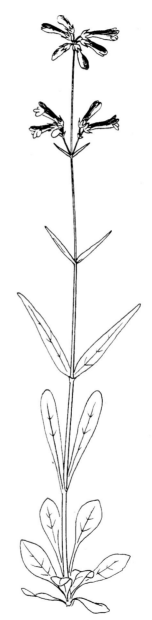

Penstemon oliganthus
Color Plates Nos. 41 & 42

Flowers: The small, .6-1" (15-26 mm.) flowers have almost straight tubes and are carried on long stems, often drooping. The throat is bearded with pale golden hairs but is white in appearance. The throat has two ridges on the lower side. The lower lip is longer than the upper. The flowers are glandular-hairy on the outside.

Leaves: The dark green leaves may be smooth or hairy and have pointed tips. The basal leaves are ovate, elliptic or lanceolate with blunt or sharp tips. They are carried on long stems and are present when blooming. The stem leaves are lanceolate or linear, shorter than the length of the stem between the leaf nodes. They grow only 2/3 up the flowering spike which gives the stem a somewhat bare appearance.

Calyx: The calyx is .16-.24" (4-6 mm.) long with broad lanceovate to lanceolate lobes, which taper to a point, with papery margins.

Stamens: The staminode is bearded with gold hairs and is clearly visible at the mouth.

PENSTEMON GRIFFINII

Penstemon griffinii A. Nelson. Bot. Gaz. 56:70 1913 Type: from Wagon Wheel Gap, Mineral Co. Colorado

Penstemon griffinii has been found in pine forests, rocky hillsides and dry grasslands of southern Colorado and Taos and Rio Arriba counties in northern New Mexico. *(See Color Plate No. 43)*

Height and Growth Habit: The plant grows 4-20" (10-50 cm.) with one to several stems that have fine hairs on the lower part.

Flowers: The slender and straight, one-sided inflorescence has no leaves and the .7-1" (17-25 mm.) blue or purple flowers are pale beneath and glandular. The throat is densely covered with golden hairs and has two ridges on the lower side.

Leaves: The leaves are smooth. The basal leaves are stemmed, spatulate and form a rosette which is present at blooming. Stem leaves are without stems, linear and are narrower toward the inflorescence.

Calyx: The calyx lobes are .18-.28" (4.5-7 mm) long, lanceolate, sharp-tipped, with papery margins. They may be moderately glandular or smooth.

Stamens: The fertile anthers are toothed along the opening. The staminode is slightly dilated at the tip and does not protrude from the throat. It is densely bearded for about half its length with stiff golden hairs.

PENSTEMON INFLATUS

Crosswhite Am. Mid. Nat. 74(2):436, 1965

Penstemon inflatus is found at 7500-11,000' elevations in the Manzano and Sandia Mountains and the mountains north to Colorado.

Height and Growth Habit: The plant has one or several microscopically hairy stems 1.5-9.6" (4-24.6 cm).

Flowers: The inflorescence is spreading, glandular to smooth. The bracts are somewhat glandular and become very small toward the top of the inflorescence. The blue flowers are .68-1" (17-27 mm), abruptly inflated, lighter below with straight guide-lines on the lower side of the throat, which is moderately ridged. No hairs are present.

Leaves: The leaves are smooth and slightly waxy. The lanceolate basal leaves are stemmed with the basal rosette often not present at flowering. The lower stem leaves are similar to the basal leaves but become narrower toward the inflorescence.

Calyx: The .12-.4" (3-10 mm.) lobes are lanceolate to narrowly linear and are glandular to smooth with age.

Stamens: The staminode is bearded with yellow-orange hairs for about 1/2 the length and does not project out of the throat.

Penstemon inflatus

PENSTEMON METCALFEI

(Metcalfe's penstemon)
Penstemon Metcalfei Woot. and Standl., Torreya 9:145. 1909 *** *Penstemon puberulus* Woot. & Standl. Bull. Torrey Club 36: 112, 1909, not M.E. Jones, 1908

Previously listed as *Penstemon whippleanus*, *Penstemon metcalfei* has recently been restudied and appears to show sufficient differences to be considered a separate species related to the Oliganthus group. Originally described from the Lookout Mine in the Black Range, it has been found also near Emory Pass at 7500' to pine-fir elevations at 9000'. It is included on the New Mexico Rare Plant Review List. *(See Color Plates Nos. 44 & 45)*

Height and Growth Habit: The stems are covered with whitish hairs and range from 5.8-17.5" (15-45 cm.).

Flowers: The pale purple flowers are .6-1" (15-25 mm.) long, glandular-hairy externally, with white hairs on the lower lip. The width at the throat is up to .24" (6 mm.). The width across the outer lobes at the lower lip is up to .48" (12 mm.).

Leaves: The leaves of the basal rosette are mostly ovate and stemmed. The stem leaves are mostly lanceolate and without stems.

Calyx: The lobes are up to .24" (6 mm.) long, tapering to a sharp tip.

Stamens: The staminode is bearded with golden hairs and is visible but not protruding.

PENSTEMON PSEUDOPARVUS

Crosswhite Am. Mid. Nat. 74(2): 431, 1965. Type from top of Mt. Withington, San Mateo Mts., Socorro Co., N.M. collected by Nisbet in flower, July 11, 1952.

Penstemon pseudoparvus occurs in grassy meadows around 10,000' in the Magdalena Mountains as well as Mount Withington. It is on the New Mexico Rare Plant Review List.

Height and Growth Habit: The plant is small, 4-11.7" (10-30 cm.). Its one to several stems are hairy to the naked eye.

Flowers: The slender and straight inflorescence is glandular with 3-10 flowers. The .4-.7" (11-17 mm.) corolla is horizontal, not drooping, moderately glandular, little expanded. The throat is wide open and ridged on the lower part. The lower lip is bearded with moderately long white hairs.

Leaves: The stemmed, spatulate to broadly elliptic basal leaves, .2-.72" (5-18 mm.) wide, .6-.24" (15-60 mm.) long, usually form a rosette which remains at flowering. The stem leaves, .03-.2" (.8-5 mm.) wide, .72-1.68" (18-42 mm.) long, are narrowly lanceolate to linear, glandular with slightly papery margins and without stems.

Penstemon pseudoparvus
Color Plates Nos. 46 & 47

Calyx: The calyx is .14-.2" (3.5-5 mm.) long, the lobes lanceolate, glandular with slightly papery margins.

Stamens: The staminode is visible and lies within the throat. It is densely bearded for 1/2 the length with stiff golden hairs.

PENSTEMON OPHIANTHUS

Penstemon ophianthus Pennell. Contr. U.S. Natl. Herb; 20 343. 1920 *** *P. jamesii* subsp. *ophianthus* Keck. Bull. Torrey Bot. Club 65:240. 1938 (M.E. Jones 5708, "near Thurber, which is close to Loa." 7000 ft. Wayne Co. Utah. 1 Aug 1894; holotype at US!; isotype at NY!) *** *P. pilosigulatus* A. Nels. Univ. Wyoming Publ. Sci. Bot 1:130, 1926 (H.C. Hanson 554, near Flagstaff, "the canadian zone and above," Coconino Co. Ariz. 6 June 1923; holotype at RM!)

Penstemon ophianthus is closely related to *Penstemon jamesii* and has been considered a subspecies. It is now listed as a separate species. It grows in sagebrush or pinyon-juniper to ponderosa pine communities. It has been found in the Sandia Mountains and ranges west of the Rio Grande to northern Arizona and to southern Utah and Colorado.

Height and Growth Habit: The one to several stems are 5-10.5" (13-27 cm.) tall and are somewhat sticky or glandular-hairy.

Flowers: The .5-.8" (14-20 mm.) lavender, violet or blue-violet flowers (rarely, white) have deep purple guide-lines on all lobes. The white throat carries guide-lines as well and is glandular hairy inside and outside and carries long, soft white hairs.

Leaves: The 1-3" (3-7.5 cm.) leaves may be slightly toothed. The basal leaves may be oblanceolate and stemmed. The stem leaves are linear to lanceolate.

Penstemon ophianthus
Color Plates Nos. 48 & 49

Calyx: The glandular-hairy calyx is .24-.4" (6-10 mm.) with narrow lanceolate lobes with sharp tips and papery edges.

Stamens: The staminode protrudes from the throat and is heavily bearded with yellow hairs.

P E N S T E M O N   P A L M E R I

(Balloon flower, Palmer's penstemon, Pink wild snapdragon)
Penstemon palmeri A. Gray. Proc. Amer. Acad. Arts 7:379. 1868 (*Coues & Palmer 228.*
"Rocky River banks. Rio Verde." Skull Valley, Ariz., 28 Aug 1865; holotype at GH!) ***
P. macranthus Eastw. Bull. Torrey Bot. Club 32:207, 1905 *** *P. palmeri* var. *macranthus*
N. Holmgren. Brittonia 31:105.1979. (O.F. Heizer s.n. IXL Canon. e. side of the
Stillwater Range. 33 airline mi. ene. of Fallon. Churchill Co. Nev., 15 June 1902; holo-
type at CAS!) *** *P. palmeri* subsp. *eglandulosus* Keck. Amer. Midl. Naturalist 18: 797.
1937 *P. palmeri* var. *eglandulosus* N. Holmgren. Brittonia 31:105, 1979 (B. Maguire, R.
Maguire & G. Piranian *12279.* "Base of red sandstone cliffs, " 2.5 mi. n. of Kanab, Kane
Co., Utah, 29 June 1935; holotype at UTC! isotype at NY!)

Penstemon palmeri can be very tall; depending on moisture, it may grow to
6'. Its puffy, pink to white blooms are very showy and appear in June or July.
It is one of the few penstemons with a noticeable fragrance. It grows on
dry, gravelly roadsides and meadows at lower altitudes in central and western New Mexico, extending into Utah and west into Arizona and Nevada
and southern California.

Height and Growth Habit: The flowers are carried on one to
several stout stems 19"-5.8' (50-180 cm.) or sometimes taller.

Flowers: The 1-1.4" (25-35 mm.) glandular-hairy flowers are
carried on one side of the stout stems. They are white, pale
pink or rose-pink, with red guide lines. The throat is abruptly
and widely expanded. The lobes are reflexed. The lower ones
are bearded with long white hairs and are twice as long as the
upper.

Leaves: The gray-green, somewhat waxy leaves may have
teeth tipped with a sharp rigid point. The basal and lower
stem leaves are lanceolate to lanceovate, their edges toothed
with sharp rigid points, and are stemmed. The upper stem
leaves are more ovate, not toothed and are joined to encircle
the stem. The bracts and upper stem leaves are not joined
around the stem and are not conspicuously toothed.

Penstemon palmeri
Color Plates Nos. 50 & 51

Calyx: The .16-.24" (4-6 mm.) lobes are ovate, pointed, downy and papery.

Stamens: The protruding hooked staminode is glandular-hairy at the base and is heavily bearded with yellow hairs on the last 1/4 of its length.

PENSTEMON PINIFOLIUS

(Pine needle penstemon)
Penstemon pinifolius Greene, Bot. Gaz. 6:218. 1881

Penstemon pinifolius is an attractive low-growing evergeen plant with showy, scarlet flowers from June to August. It grows in limestone and gravel in the rocks and cliffs of the high mountains of southern New Mexico, (common in the Magdalena and San Mateo Mountains and the Black Range) and is found also in southeastern Arizona and nearby Mexico. A yellow strain has been found from time to time in the Magdalena Mountains.

Height and Growth Habit: The low mats send up numerous flowering stems to 8"- 9.75" (20-25 cm.).

Flowers: The long 1-1.28" (25-32 mm.), narrow, scarlet flowers stand well above the foliage, carried on one side of the glandular-hairy stems. The lower lip is widely divided into 3 flaring long, narrow petals which carry scattered yellow or white hairs. The flower stems and outer surface of the flowers are glandular-hairy.

Leaves: The thick .24-.8" (6-20 mm.) smooth leaves are dark green, needle-like, narrow, crowded at the base of the stem, scattered on the blooming stalk.

Calyx: .2-.28" (5-7 mm.) lobes are lanceolate or long and narrow, sharply pointed, and papery at the base of the lobes.

Penstemon pinifolius
Color Plates Nos. 52 & 53

Stamens: The staminode lies within the throat, is not dilated at the tip and is bearded with bright yellow hairs for most of its length. The very small anthers project almost to the tip of the upper petals.

PENSTEMON PSEUDOSPECTABILIS

(Perfoliate penstemon, Desert penstemon)
Penstemon pseudospectabilis M. E. Jones ssp. *connatifolius* (A. Nels.) Keck, Amer. Midl. Nat. 18:807. 1937 *** *Penstemon spectabilis* Woot. and Standl., not of Thurb., Contr. U.S. Nat. Herb. 19:583. 1915 *** *Penstemon connatifolius* A. Nels., Amer. Jour. Bot. 18:437. 1931.

A large, sometimes sprawling plant with broadly spreading flower clusters of many deep rose-pink, tubular flowers on much divided stems. It blooms in April and May to June at higher altitudes in the pinyon-juniper to yellow pine elevations of southwestern New Mexico and Arizona.

Height and Growth Habit: The smooth stalks can be much divided and arise from a woody base, 16-39" (40-100 cm.) tall. The side flowering stems are often reddish in color.

Flowers: The .88-1.3" (22-33 mm.) flowers are a rich pink or magenta, on fairly long flowering stems. The lip has a few scattered hairs. The tubes are glandular inside and out. Guidelines may be present.

Leaves: The leaves are smooth, deep green or grayish-green, usually but not always markedly toothed at the edges. The basal leaves are stemmed, lanceovate to broadly ovate. The upper stem leaves may join around the stem and may reach 7" (18 cm.) across the two leaves.

Calyx: The short, .2-.28" (5-7 mm.) sepals are ovate to elliptic, sharp-tipped, with narrow papery margins.

Stamens: The narrow or somewhat dilated staminode lies within the throat and is smooth or occasionally has a few scattered hairs.

Penstemon pseudospectibilis
Color Plates Nos. 54 & 55

PENSTEMON PULCHELLUS

Penstemon pulchellus Lindl., Edwards Bot. Reg. 14: pl. 1138, 1828.

This penstemon has not been found in recent years and its existence in New Mexico is doubtful. It was collected in the San Luis Mountains probably in 1890. It is thought to have a violet corolla with the underpart very pale.

Height and Growth Habit: The stems are medium in height, hairy, and somewhat woody at the base.

Flowers: The flowers are .8-1" (20-25 mm.), slightly glandular-hairy externally. The throat is inflated and lightly haired at the base of the lower lobes.

Leaves: The leaves are smooth, the large ones are lanceolate or oblong, acute, definitely toothed; the smaller are obscurely toothed and are born in clusters in the axil of each larger leaf.

Stamens: The staminode is dilated and bearded at the tip with a tuft of short yellows hairs.

PENSTEMON RAMOSUS

Penstemon ramosus Crosswhite SIDA 2:339-346 *** *Penstemon pauciflorus* Greene, in Bot. Gaz. 6:218. 1888. Type: "Bluffs of the Rio Gila", August 30 1880, E.L. Greene. By the time Greene finally named the species he had probably already distributed the species under another manuscript name which proved to be a later homonym, necessitating the change.

The differences between this species and *Penstemon lanceolatus* are so subtle as to have caused some confusion. *Penstemon lanceolatus* is a Mexican species that now is considered to come into New Mexico only at the Mexican border, if at all. Its brilliant scarlet tubes appear in summer and are carried on slender upright stems with gray leaves so narrow the plant almost disappears when not in bloom. The most recent study from herbarium samples by Dr. Frank Crosswhite states that both have red tubular corollas but *P. lanceolatus* is unbranched below the inflorescence, with leaves that are .16-.32" (4-8 mm.) wide and do not turn back on the edges. He indicates that *P. ramosus* differs from *P. lanceolatus* by its habit of branching below the inflorescence, leaves that turn back on the edges, the leaves of the branches .04" (1 mm.) wide and those of the stem .12-.24" (3-6 mm.) wide.

The openings of the anthers of *P. lanceolatus* are described as being toothed at the edges, while those of *P. ramosus* are not. The very small scattered populations of *P. ramosus* occur in dry, rocky areas of southern New Mexico and Arizona. It is on the New Mexico Rare and Sensitive Plant Species List.*(See Color Plates Nos. 56 & 57)*

Height and Growth Habit: The hairy stems 11.7-29" (30-75 cm.) arise from a woody rootstock and are often branching below the inflorescence.

Flowers: The narrow inflorescence is covered with gland-tipped hairs with only one branch at a node bearing one or two flowers. The 1.1-1.6" (28-40 mm.) corollas are straight.

Leaves: The very narrow, linear-lanceolate, leaves are 2.34-4.29" (6-11 cm.) on the main stem. They are somewhat hairy and their edges turn upward toward the center. The leaves of the branching stems are about 1" (25 mm.), sometimes longer.

Calyx: The glandular sepals are .2-.4" (5-10 mm.).

Stamens: The fertile anthers are twisted, the openings not toothed. The staminode is smooth, lies within the throat and is not dilated.

PENSTEMON ROSTRIFLORUS

(Beak-flowered penstemon, Bridge's penstemon)
Penstemon rostriflorus Kellogg. Hutchings' California Magazine 5(3): 102. 1860: Proc. Calif. Acad. Sci. 2:15 1863. *** *P. bridgesii rostriflorus* Schelle in Beissner, Schelle & Zabel. Handb. Laubholzben 432. 1903. (*Mr. J. M. Hutchings s.n.* "crevice of the Lower Dome at the back of the Great Tissaac, or South Dome—3500 feet above the Yosemite Valley." Mariposa Co., Calif.) According to Curran (Bull. Calif. Acad. Sci. 1:145, 1885), the type was a fragment at CAS and resembled *P. bridgesii*. The specimen was apparently destroyed in the San Francisco earthquake, but the figure in the protologue is adequate as the type. Kellogg's description is unmistakable for this taxon, the only divergence being the "creamy yellow" flowers, a possible variation from the normal orangish-red. *** *P. bridgesii* A. Gray, Proc. Amer. Acad. Arts 7:379, 1868. (*Bridges 218*, "Collected in California" holotype at GH! isotype at NY!) *** *P. bridgesii* var. *amplexicaulis* Monnet. Bull. Soc. Bot. France 61: 228, 1914, (*Monnet 1036*, "Dans les washes sablonneux," Gold Mt. 1850 m. Esmeralda Co., Nev.)

Formerly called *Penstemon bridgesii*, *Penstemon rostriflorus* is an open, many stemmed plant with scarlet flowers from June to September. The horseshoe-shaped anthers that open across the top distinguish it from other red penstemons. It grows in dry washes and cliffs at pinyon and ponderosa elevations of western New Mexico and ranges west to California, southern Nevada and Utah, and into southern Colorado and northern Arizona.

Height and Growth Habit: The bushy plant grows from about 12-23" (30-60 cm.), with many leaves closely placed on the branches near a woody base. The stems may be smooth or hairy.

Flowers: The .88-1" (22-27 mm.) scarlet flowers are glandular on the outside and are carried on erect, nearly one-sided, glandular-hairy stems. The flower face is narrow (narrower than *P. barbatus*, which it resembles at first glance) with a projecting hooded or beaked upper lip separated only at the very tip. The lower lip curves back sharply into three long, widely separated petals. The buds are greenish-yellow, scarlet tipped. There are four or more flowers to the side stems, often 8 or 9 flower clusters to a stem. The flower tube has 2 shallow ridges on the lower side.

Leaves: The leaves are smooth, and dark green. The stemmed lower leaves are narrow oblanceolate, the upper stem leaves linear, widely separated on the stem.

Calyx: The calyx is .16-.24" (4-6 mm.) long, glandular, lanceolate to ovate, tips sharp and often recurved and may have papery edges near the base.

Stamens: The staminode is smooth, and lies inside the throat. The yellowish anthers are horseshoe-shaped, opening across the top and minutely toothed at the opening. They reach almost to the tip of the upper lobe.

Penstemon rostriflorus
Color Plates Nos. 58 & 59

PENSTEMON RYDBERGII

(Rydberg's penstemon, Meadow penstemon)
Penstemon Rydbergii A. Nels., Bull. Torr. Bot. Club 25:281. 1898. *** *Penstemon erosus*
Rydb., Bull. Torr. Bot. Club 28:28. 1901 *** *Penstemon lacerellus* Greene, Leaflets 1:161.
1906.

Penstemon rydbergii is one of the few penstemons that is fond of damp
places in high mountain meadows or swamps close to streams. It grows in
colonies that make a brilliant show of color. The small flowers of intense
color, with blue-purple throats and deep blue faces, are carried in many
tight whorls around the stem. All the sepals and flower bracts have wide
papery, ragged edges. Blooming time is July and August in the mountains
of northern New Mexico above 7000'.

Height and Growth Habit: The plant grows
from 8-27" (20-70 cm.). Two kinds of stems
are usually present: slender, erect flowering
stems and short leafy stems without flowers,
both coming up from a branched woody base.
The stems may be smooth or hairy, often red-
dening toward the tip.

Flowers: The small, .4-.56" (10-14 mm.)
(sometimes up to .75") flowers have bright to
deep blue faces with purple tubes, or are all
purple. The narrow or very slightly expanded
tubes are smooth on the outside. The flower
face has a light-colored ridge across the lower
lip and a light-colored throat, which is heavily
furred with golden hairs at the opening. The
flowers and bracts are crowded in the widely
separated clusters around the upper part of the
stem. The bracts occur among the flower clusters and are papery and ragged at their edges.

Leaves: The leaves are dark green. The leaves
of the non-flowering shoots or basal leaves are
usually smooth, oblong to oblanceolate, blunt
or coming to a point, often clasping the stems.
The stem leaves on flowering stems are few,
smooth, oblong to oblanceolate, come to a
point and occur in pairs along the stem.

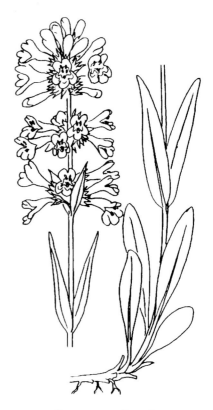

Penstemon rydbergii
Color Plates Nos. 60 & 61

Calyx: The lobes are .16-.2" (4-5 mm.), sharp-tipped, with broad, papery, ragged edges.

Stamens: The staminode is narrow with a yellow beard at its tip.

PENSTEMON SECUNDIFLORUS

(One-sided penstemon, Side-bells penstemon)
Penstemon secundiflorus Benth., DC. Prod. 10:325. 1846

Penstemon secundiflorus is one of the earliest blooming penstemons in grasslands and foothills of central and northern New Mexico, blooming in late April or early May to June. Its flaring flowers are variable in color, blue or violet, red-violet, lilac or even pink or turquoise, rarely white. The large staminode, heavily bearded with yellow hairs, is clearly visible.

Height and Growth Habit: 6-19.5" (15-50 cm.) tall, the plant is smooth and somewhat waxy.

Flowers: The flowers are up to 1" (17-25 mm.) and the corollas gradually expand. The color is variable and can be medium blue, violet, red-violet, lilac or pink or turquoise, or rarely white. The petals spread from a flat plane. Hairs are present on the lower lip. Guidelines are sometimes visible on the upper lobes but usually are obscured or not present on the lower lobes.

Leaves: Dark to waxy grey- green, the stem leaves are often erect, lanceolate to lanceovate with pointed tips. The basal leaves form a rosette and are obovate to spatulate, stemmed and blunt.

Penstemon secundiflorus
Color Plates Nos. 62 & 63

Calyx: The lobes are .16-.28" (4-7mm.), sharp at the tips and have papery edges.

Stamens: The dilated staminode is heavily bearded with yellow hairs.

PENSTEMON STRICTIFORMIS

Penstemon strictformis Rydb. Bull. Torrey Bot. Club 31:642. 1904 (1905) *** *P. strictus* Benth. subsp. *strictiformis* (Rydb.) Keck J. Wash. Acad. Sci. 29:491, 1939 (*Baker, Earl & Tracy 76*, "locally common on foothills near town," Mancos, 7200 ft. Montezuma Co., Colo., 23 June 1898; holotype at NY!)

Penstemon strictiformis lives up to its name in that the blooming stems are very stiffly upright, although some may be slightly curved toward the woody base. The anthers that have long tangled hairs and the flower form and color may be reminiscent of *P. strictus* but it is considered a separate species. It is found in northwestern New Mexico, Utah, Colorado and Arizona at juniper and pinyon-juniper elevations, blooming in late May to July.*(See Color Plates Nos. 64 & 65)*

Height and Growth Habit: One to several stems, 7.8-21" (20-55 cm.), arise from a woody base. All parts of the plant are smooth.

Flowers: The 1-1.4" (25-35 mm.) blue-lavender flowers usually are carried on one side of the stem. They are somewhat swollen on the under side but almost flat on the upper side, smooth inside and out.

Leaves: The 2.3-3.5" (6-9 cm.) basal and lower stem leaves are narrowly oblanceolate and taper to a stemmed base. The 1.5-3" (4-8 cm.) upper stem leaves are lanceolate and stemless.

Calyx: The .24-.32" or up to .38" (6-8 or up to 9.5 mm.) lobes are lanceolate to ovate, sometimes with a sharp point at the tip. The lower edges are ragged and papery.

Stamens: The smooth or lightly bearded staminode is dilated and may curve up at the tip. The fertile anthers extend out of the throat and are covered with long, tangled hairs, their openings somewhat toothed.

PENSTEMON STRICTUS

(Rocky Mountain penstemon, Porch penstemon, Stiff beardtongue)
Penstemon strictus Benth. in A. DC. Prodr. 10:324, 1846. (Fremont, in montibus Scopulosis ad fontes fl. Sweetwater, South Pass, Sweetwater Co., Wyo. 7 Aug. 1842; at NY!) *** *P. strictus* subsp. *angustus* Pennell, Contr. U.S. Natl.. Herb. 20:356. 1920 (*C.F. Baker 604*, Piedra, Archuleta Co., Colo. July 1899; holotype at NY!) A puberulent form.

Penstemon strictus is a variable species, seeds itself freely and crosses easily with many other penstemons. It often forms large colonies in open woods and meadows or along roadsides from pinyon-juniper zones up to the high mountain spruce-aspen elevations in northern and central New Mexico. Its showy blooms of intense blue or purple appear from June to August. Its anthers are visible in the throat and are covered with short, fuzzy whitish hairs which darken with age.

Height and Growth Habit: The many erect, smooth, strong stems, 7.8-31" (20-80 cm.), arise from a woody rootstock.

Flowers: The .7-1.1" (18-28 mm.) flowers are carried on one-sided, erect flowering stems. Their color varies from purple to a striking blue with purple in the throat and tube but the overall impression is of an intense blue, many plants carrying flowers of both colors in each blossom.

Leaves: The leaves are dark green, smooth and rounded at the tip. The basal leaves are oblanceolate or spatulate, usually with long stems. The stem leaves are linear to broadly lanceolate.

Calyx: The thin lobes are ovate to lanceolate .12-.24" (3-6 mm.) with papery edges pointed at the tip.

Stamens: The smooth or sparsely bearded staminode lies within the throat and has a slightly dilated tip. The anthers have long fuzzy whitish hairs and are visible in the throat.

Penstemon strictus
Color Plates Nos. 66 & 67

PENSTEMON SUPERBUS

(Superb penstemon)

Penstemon superbus A. Nels., Proc. Biol. Soc. Wash. 17:100. 1904. *** *Penstemon puniceus* A. Gray, U.S. and Mex. Bound. Bot. Rpt. 113. 1859. Not *P. puniceus* Lilja, 1843

The brilliant coral-red flowers of *Penstemon superbus* produce one of the most beautiful sights of the hot, dry deserts of southwestern New Mexico, eastern Arizona and northern Mexico from April to June. It is an erect plant and can be very tall, sometimes reaching 3 or 4'. It is on the New Mexico Rare and Sensitive Plant Species List.

Height and Growth Habit: The long smooth and waxy stems may reach 4' (30-120 cm.) or taller, often producing blooming side shoots. They may be red-purple.

Flowers: Its many .68-.88" (17-22 mm.) striking coral-pink to red trumpet-shaped flowers bloom on all sides of the stems, which may be branching. The flowers carry glandular hairs both outside and on the lobes. The flat flower face is almost round and only obscurely 2-lipped.

Leaves: The grayish leaves are pointed or sometimes blunt-tipped and clasp the flower stem or can be almost joined around the stem. The basal leaves are stemmed, oblanceolate, spatulate or elliptic. The thick stem leaves are broadly ovate, elliptic or oblong-ovate, pointed at the tip and clasping or tending to surround the stem.

Calyx: The .16-.2" (4-5 mm.) lobes are ovate or elliptic, pointed at the tip, with narrow papery edges.

Stamens: The staminode shows in the throat but does not project out. It has golden hairs on the upper half. Its dark red tip is slightly flattened and slightly notched but not dilated.

Penstemon superbus
Color Plates Nos. 68 & 69

PENSTEMON THURBERI

(Thurber's penstemon)

Penstemon thurberi Torr., U.S. Rpt. Expl. Miss. Pacif. 7:15. 1856 (*P. scoparius* A. Nels. from West Wells *Gooding 1037*) ***Leiostemon thurberi* Greene, Leaflets 1:223. 1906.

Penstemon thurberi grows in the dry grasslands of south central and southwestern New Mexico. Its range extends into southern California, southern Arizona and northern Mexico. Its delicate pink flowers appear from April to August. It is easily confused with *P. ambiguus* but *P. thurberi* has a deeper colored flower face and the throat opening is more wide open while the tube remains narrow. The flower face of *P. ambiguus* is white and the opening to the throat is narrow, almost closed.

Height and Growth Habit: The shrubby, many branched plant grows from 8-16" (20-40 cm.) from a woody base.

Flowers: The .4-.56" (10-14 mm.) flowers are dark pink with a blue cast, usually carried singly. (Blue populations have been reported.) The straight throat expands very slightly at the face. The flower is slightly downy at the base of the lower lobes, occasionally around the opening to the throat and two rows of hairs extend down the throat.

Leaves: The smooth leaves are very narrow, linear, .25" up to 1" (5-25 mm.) long, sharp-tipped.

Calyx: The tiny .08" (2 mm.) ovate pointed lobes have slightly papery edges.

Stamens: The fertile anthers are very small. The staminode is very narrow and smooth.

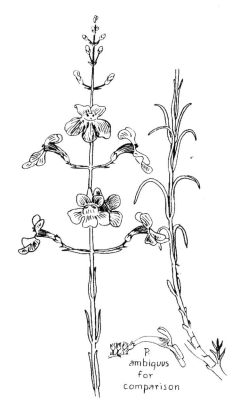

P. ambiguus for comparison

Penstemon thurberi
Color Plates Nos. 70 & 71

PENSTEMON VIRGATUS

(Wandbloom penstemon, Varied penstemon)
Penstemon virgatus A. Gray, U.S. and Mex. Bound. Bot. Rpt. 113. 1859. The type was collected near Santa Rita by Bigelow and Wright in 1851.

From June to August or September, depending on moisture, the flowers of *Penstemon virgatus* appear on long, wand-like one-sided stalks throughout the mountain meadows and pine woods in north central to southern and western New Mexico. The puffy flowers are often pale and variable in color.

Height and Growth Habit: The 9.75-31" (25-80 cm.) or taller, single or many, downy or smooth stems are straight and slender.

Flowers: The .6-1" (15-24 mm.) flowers are pallid white, pale lavender, blue or occasionally pink, with red-purple guidelines on the lower lobe. The throat is smooth on the outside, swollen both top and bottom. The lobes project or spread, the lower ones may be reflexed and may carry a few hairs. The inflorescence is one-sided.

Leaves: The narrow leaves are .78-4.7" (2-12 cm.), linear or lanceolate, smooth or slightly hairy.

Calyx: The sepals are short .12-.16" (3-4 mm.) and broad, ovate, elliptic or obovate with blunt or pointed tips. They are papery and somewhat ragged at the edges.

Stamens: The smooth, narrow staminode may be slightly dilated.

Penstemon virgatus
Color Plates Nos. 72 & 73

PENSTEMON WHIPPLEANUS

(Whipple's penstemon, Dusky penstemon)

Penstemon whippleanus A. Gray, Proc. Amer. Acad. Arts and Sci. 6:73. 1862. The type was collected in an arroyo in the Sandia Mountains east of the Rio Grande by Bigelow in 1853 *** *Penstemon glaucous* var. *stenosepalus* A. Gray, Proc. Amer. Acad. Arts and Sci. 6:70. 1862 *** *Penstemon arizonicus* Heller, Bull. Torr. Bot. Club 26:591. 1899. *** *Penstemon stenosepalus* (A. Gray) Howell, Fl. Northw. Amer. 1:514. 1901.

Penstemon whippleanus is a high mountain dweller with very dark, blue or red-purple flowers from June to August. It is found in Wyoming, Idaho, Montana, Colorado, Utah and Arizona as well, where the flower color is sometimes much lighter yellow or white, brownish, or blue or purple.

Height and Growth Habit: The 5.8-23" (15-60 cm.) smooth or hairy stems are few to several, slender and arise from horizontal woody rootstocks. The entire plant may be downy.

Flowers: At .88- 1.2" (22-30 mm.), the noticeably glandular hairy flowers often droop and are in widely separately clusters or crowded at the top of the stem. The color is usually very dark, often somber, varying from dark blue or wine to dark purple, sometimes almost metallic in appearance. The interior of the lip is light in color and dark guidelines lines appear on the tube and throat. The projecting lower lip carries long white hairs.

Leaves: The leaves are dark green and may be slightly toothed along the edges. The basal leaves are ovate to spatulate, often long-stemmed. The stem leaves are lanceolate to oblanceolate, pointed and without stems. The stemless bracts appear only at the base of the flower cluster.

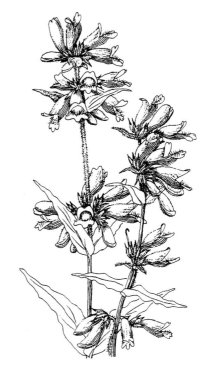

Penstemon whippleanus
Color Plate No. 74

Calyx: The very long, .28-.4" (7-10 mm.), narrow, pointed sepals are glandular-hairy with narrow papery edges at the base.

Stamens: The staminode is smooth or has a few gold hairs at the tip, which may be slightly dilated.

PENSTEMONS IN THE GARDEN

In recent years, penstemons have become a beautiful addition to our gardens. Now nurseries carry the started plants and seed can be obtained through a number of sources. Among my favorites for easy growing in New Mexico are *P. crandallii, pinifolius* and *linarioides* for ground cover. For the perennial garden, I like the showy and taller *P. angustifolius, eatonii, barbatus, pseudospectabilis, palmeri,* and, the most reliable of all, the beautiful Rocky Mountain penstemon, *P. strictus.*

Requirements: Though many penstemons are tolerant of varying conditions, for best success in the garden, match the kinds with your location — mountain species with cooler, higher locations, those from the desert with warm, dry gardens. Penstemons demand good drainage and full sun. Overwatering and soggy soil bring on an early death. New Mexican penstemons, native to alkaline soils, do best in gravelly, sandy soils and scree. A few, like *P. strictus*, are more tolerant of heavier soils. Once established, they require a minimum of care. In the wild they are true pioneers, seeding themselves happily among rocks and gravel of disturbed ground and roadsides.

Growing Penstemons from Seed: Growing penstemons from seed is not difficult but requires some special conditions. Gather seeds in the fall and store for at least six weeks in cool, dry conditions. If the capsules are not open, store them in a paper bag inside in a garage through the winter. Early the next year, shake the stems inside the bag and most of the seeds will fall out. If they don't, the capsules may have to be broken up with a hammer. If seed is gathered from the garden remember that penstemons cross easily; you may harvest some exciting hybrids. In the wild, take only a few of any wild population to ensure their survival. Seeds of rare or sensitive species should not be gathered.

Starting plants inside: In late January or February, the seed, along with a handful of slightly dampened vermiculite, can be put in sandwich bags which then are tightly closed and refrigerated for 4-6 weeks. Sometimes they will germinate in the bags in the refrigerator. After the time in the refrigerator ("stratification"), they can be sprinkled on top of a mix of 1/3 peat and vermiculite mixed with 2/3 sand and covered with a layer of sand or fine gravel. Four inch pots may be used, or starting containers deep enough for the very long roots they develop even when small. Soak the containers from underneath after planting and move to light. When the seedlings have developed four or more leaves and the weather is mild, sink the pots in sand outside. In New Mexico, they need to be protected from the sun and wind after transplanting to the open ground until they are well established. If they don't germinate on the first try, sink the pots in sand outside and have patience; often they will come up the second year or even later. Penstemon seeds seem to last; even old seed often will germinate. Some desert varieties, such as *P. ambiguus and P. angustifolius*, germinate only at warm temperatures (70 degrees) and need no stratification.

Growing seeds outside: Some of my best success in growing from seeds has been to sow them directly in place in the fall, ideally in gravel or rocky soil, and leaving them to their own devices through the winter. This has been particularly successful with the obliging *P. strictus, P. angustifolius, P. eatonii* and *P. liniarioides.*

Cuttings: Softwood cuttings work well for *P. crandalli, P. pinifolius* and similar plants that have woody stems. The cuttings are dipped in rooting hormone and stuck in dampened sand or perlite and kept moist until rooted.

Pests: Penstemons are remarkably free from pests when grown in dry, well drained conditions. Grown in large numbers, they may be subject to some pests. Aphids may attack new growth but can be removed with a soapy spray. Occasionally, mildew appears on *P. strictus;* however it does not seem to harm the plants. Black spot fungus can be treated with a fungicide drench. Pittosporum pit scale can produce deformed, swollen and twisted stems. To treat, swab with alcohol, dispose of infected parts, and use a systemic insecticide. Treat in early spring with general insecticide if you have the problem.

For more complete information about propagation, see Judith Phillip's *Plants for Natural Gardens.*

R E F E R E N C E S

American Penstemon Society, *Bulletins*, 1569 South Holland Court, Lakewood, Colorado, 80226, USA

Crosswhite, Frank S., "Revision of Penstemon Section Chamaeleon (Scrophulariaceae)", *SIDA* 2:339.346

——————— "Revision of Penstemon II", *The American Midland Naturalist*, 74(2) 437-441

Harrington, H. D., How to Identify Plants, Illustrated by L. W. Durrell, Swallow Press, Athens, Ohio 1957

Kearney, Thomas H., Robert H. Peebles, et al. 1951. *Arizona Flora*, Berkeley, Calif., University of California Press

Martin, William C. and C.R. Hutchin, 1980. *A Flora of New Mexico*, J. Cramer, Germany

New York Botanical Garden, 1984. *Intermountain Flora*, Bronx, New York

Nisbet, Gladys T. and R. C. Jackson, "The Genus Penstemon in New Mexico", *University of Kansas Science Bulletin*, Vol. XLI, No. 5, December 23, 1960

Phillips, Judith, 1995, *Plants for Natural Gardens*, Museum of New Mexico Press, Santa Fe, New Mexico

Rickett, Harold William, *Wild Flowers of the United States*, New York Botanical Garden, New York, N.Y. Vols. 4,5 and 6

State of New Mexico, August 1995. *Inventory of Rare and Endangered Plants of New Mexico*, Robert Sivinski and Karen Lightfoot, eds., New Mexico Forestry and Resources Conservation Div., Energy, Minerals and Natural Resources Dept., Santa Fe, New Mexico.

Tidestrom, Ivan and Sister Teresita Kittell, 1941. *Flora of Arizona and New Mexico*, Catholic University of America Press, Washington, D.C.

Index
Common Names

Index of Latin Names

(Bold type indicates names in current use.)